Nature's Mysteries

POLAND'S CROOKED FOREST

by Rachel Hamby

abdobooks.com

Published by Pop!, a division of ABDO, PO Box 398166, Minneapolis, Minnesota 55439.

Printed in the United States of America, North Mankato, Minnesota.

102020
012021

THIS BOOK CONTAINS RECYCLED MATERIALS

Cover Photo: Shutterstock Images
Interior Photos: Shutterstock Images, 1, 6–7, 8, 9, 12, 15, 17, 19, 26, 27, 29; Christina Falkenberg/Westend61 GmbH/Alamy, 5; Konrad Zelazowski/Alamy, 11, 21; AP Images, 20; iStockphoto, 13, 23, 25

Editor: Alyssa Krekelberg
Series Designers: Candice Keimig, Victoria Bates, and Laura Graphenteen

Library of Congress Control Number: 2020940289

Publisher's Cataloging-in-Publication Data

Names: Hamby, Rachel, author.

Title: Poland's Crooked Forest / by Rachel Hamby

Description: Minneapolis, Minnesota : POP!, 2021 | Series: Nature's mysteries | Includes online resources and index

Identifiers: ISBN 9781532169229 (lib. bdg.) | ISBN 9781532169588 (ebook)

Subjects: LCSH: Plant mutation--Juvenile literature. | Forest ecology--Juvenile literature. | Curiosities and wonders--Juvenile literature. | Mystery--Juvenile literature. | Geography--Juvenile literature.

Classification: DDC 910.02--dc23

WELCOME TO DiscoverRoo!

Pop open this book and you'll find QR codes loaded with information, so you can learn even more!

Scan this code* and others like it while you read, or visit the website below to make this book pop!

popbooksonline.com/crooked-forest

*Scanning QR codes requires a web-enabled smart device with a QR code reader app and a camera.

TABLE OF CONTENTS

CHAPTER 1

THE CROOKED FOREST

Near Gryfino, Poland, there's a walking trail. Visitors follow it through a forest. At first, everything looks normal. Trees grow toward the sky. The trunks are straight. But then, things change.

WATCH A VIDEO HERE!

The pine trees in the Crooked Forest have strangely shaped trunks.

In the Crooked Forest, some trees have unusual curves.

The trunks curve between 3 and 9 feet (0.9 and 2.7 m) to the side.

The forest has many crooked pine trees. Each tree trunk has a C-shaped curve. But the plants are healthy despite their strange shape.

People think the trees were planted around 1930. But they were not discovered until the 1970s. People were surprised that so many curved trees were in one area.

FUN FACT

The Crooked Forest was found when a power plant was built nearby.

The curved trees are surrounded by straight ones. No one knows why only some trees are bent. But there are several **hypotheses** about what might have happened.

People study trees to learn more about them and the areas where they grow.

Many people are curious
about this strange forest.

CHAPTER 2

AN UNSOLVED MYSTERY

Scientists studied the crooked trees. They discovered something. When the trees were approximately seven to ten years old, something damaged the trunks. It is possible that the crooked trees were cut or pushed down.

LEARN MORE HERE!

Some tree trunks are more bent than others.

Some people wonder whether heavy snow is to blame. Young trees have thin trunks. They bend easily. In Poland,

Heavy snow can weigh down a young tree.

Saplings are small, young trees that can be easily damaged.

it snows in winter and early spring. Heavy snow may have pushed the **saplings** down. New **shoots** may have grown from the sideways trunks.

However, this idea has some problems. Many trees grow in the forest. Snow would have covered all of them. If snow bent the trunks, all the trees should be curved. But only some of them are. So, snow is probably not the reason for the trees' strange shape.

FUN FACT

The curved parts of the crooked trees all point north.

Strong winds can cause a tree to take on a curved shape.

CHAPTER 3

UNUSUAL SHAPES

Many people think that nearby farmers bent the trees on purpose. In the 1930s, people sometimes did this. Curved trees create curved wood. And curved wood could help people make certain things.

COMPLETE AN ACTIVITY HERE!

For example, some people used the wood to build boats.

People still bend wood today to make things.

The Crooked Forest may have been a tree farm. The Baltic Sea is north of the forest. Farmers may have shaped the trees for building boats. Boat knees are made with curved logs. These ship parts support the boat's structure.

FUN FACT

People also shaped trees to make furniture and instruments.

WHERE IS THE CROOKED FOREST?

Many places in Poland were destroyed during World War II.

Some people think farmers began curving the trees but never used the wood. World War II broke out in 1939.

It lasted until 1945. During this time, Gryfino was mostly ruined. The farmers may have abandoned their work with the trees.

Today, more than 20,000 people live in Gryfino.

PARTS OF A TREE

LEAVES

Leaves use sunlight to make sugar. This is food for the tree.

TRUNK

Water and nutrients travel up the trunk. They move through a layer under the bark.

ROOTS

Roots absorb water and nutrients from the soil.

leaves
trunk
nutrients and water
roots

CHAPTER 4

DAMAGED TREES SURVIVE

People aren't sure why the trees in the Crooked Forest got their curved shape. But trees grow toward sunlight. They need it to survive. So the bent trees eventually started to grow straight again.

LEARN MORE HERE!

Trees need water, sunlight, and nutrients to grow.

TREE RINGS

Scientists can study tree rings to learn how trees grow. Tree rings show how old a tree is. A new ring forms every year. The width between the rings changes depending on the environment. Wide rings form during rainy years. Narrow rings form during dry years. If a tree is damaged, a scar will show in the rings.

Tree rings can tell scientists what the area's climate was like years ago.

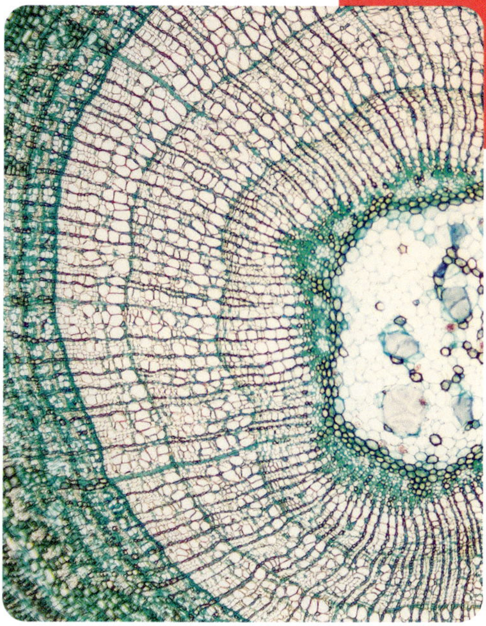

Scientists can look at plant cells with a microscope.

The crooked trees also started to grow straight because their plant cells told them to. The trees were able to **redirect** their trunks upward. This could have happened once the pressure from whatever was bending them went away.

Poland's crooked trees have been growing for 90 years. There are many ideas about the trees in the forest. The reason for their unusual shape is a mystery. They may have been pushed down, cut, or bent. But their curved trunks didn't stop them from growing upward toward the sun.

FUN FACT

The crooked trees are about 50 feet (15 m) tall.

The Crooked Forest is called "Krzywy Las" (SHIV-a LAS) in Poland.

MAKING CONNECTIONS

TEXT-TO-SELF

Would you want to visit the Crooked Forest? Why or why not?

TEXT-TO-TEXT

Have you read other books about trees? How are Poland's crooked trees similar and different?

TEXT-TO-WORLD

People don't know for sure what caused the trees' trunks to curve. What are some other mysteries that scientists haven't solved?

GLOSSARY

hypothesis — a proposed idea that is based on little evidence and needs to be studied more.

power plant — a station that creates electrical power for use in homes and businesses.

redirect — to change the direction of something.

sapling — a young tree with a thin trunk.

shoot — a new growth on a plant or tree.

INDEX

ONLINE RESOURCES

popbooksonline.com

Scan this code* and others like it while you read, or visit the website below to make this book pop!

popbooksonline.com/crooked-forest

*Scanning QR codes requires a web-enabled smart device with a QR code reader app and a camera.